The Newport Naval Training Station

A Postcard History

Federico Santi

Schiffer Publishing Ltd
4880 Lower Valley Road • Atglen, PA 19310

Acknowledgments

Thanks to The Newport Naval War College Museum and many Newport residents who have been interviewed over the years remembering the Navy Personnel from the Training Station.

Note that language viewed on postcards or other written correspondence are intentionally left as originally created.

Other Schiffer Books By The Author:

Zsolnay Ceramics: Collecting a Culture
ISBN: 0-7643-0534-4 $49.95

Newport Mansions: Postcards of the Gilded Age
(with John Gacher)
ISBN: 0-7643-2497-7 $24.95

Art Nouveau Ironwork of Austria & Hungary
(with John Gacher)
ISBN: 0-7643-2436-5 $49.95

Library of Congress Control Number: 2013930176

Designed by Mark David Bowyer
Type set in Schneidler Blk BT / Zurich BT

ISBN: 978-0-7643-4323-0
Printed in China

Schiffer Books are available at special discounts for bulk purchases for sales promotions or premiums. Special editions, including personalized covers, corporate imprints, and excerpts can be created in large quantities for special needs. For more information contact the publisher:

Published by Schiffer Publishing, Ltd.
4880 Lower Valley Road
Atglen, PA 19310
Phone: (610) 593-1777; Fax: (610) 593-2002
E-mail: Info@schifferbooks.com

For the largest selection of fine reference books on this and related subjects, please visit our website at
www.schifferbooks.com
We are always looking for people to write books on new and related subjects. If you have an idea for a book, please contact us at
proposals@schifferbooks.com

This book may be purchased from the publisher.
Please try your bookstore first.
You may write for a free catalog.

In Europe, Schiffer books are distributed by
Bushwood Books
6 Marksbury Ave.
Kew Gardens
Surrey TW9 4JF England
Phone: 44 (0) 20 8392 8585; Fax: 44 (0) 20 8392 9876
E-mail: info@bushwoodbooks.co.uk
Website: www.bushwoodbooks.co.uk

Contents

If you have a Newport postcard question or wish to contact the author, please send a self-addressed stamped envelope to:

The Drawing Room
152 Spring Street
Newport, RI 02840

You may also phone 401-841-5060 or visit online at www.drawrm.com.

Dedication

This book is dedicated to the men and women who have entered the Newport Naval Training Station and gone on to serve as sailors in the United States Navy.

Introduction
A Short History

Nicholas Easton first viewed what was to be called Coasters Harbor Island in April 1639. Benedict Arnold and John Greene of Newport purchased the island from Cojnoquant, Chief of the Sachem Indians of Narragansett Bay on May 22, 1658. They paid 6 pound 19 shillings for the land which was the equivalent of about $30. On May 1, 1673, Arnold and Greene sold the island to the City of Newport for about 10 pounds. The city used Coasters Harbor Island for a variety of purposes including the site of the Newport Asylum for the Poor established in 1820. (That structure eventually served as the Headquarters for the Naval Training Station and today is used to house the Naval War College Museum). Many naval scholars believe the first naval battle of the Revolutionary War took place at the Coasters Harbor Island in 1775 between the Colonial sloop *Katy* and the British sloop *Diana,* a tender to the British frigate *Rose*. Coasters Harbor Island contains about 90 acres and is about 2 miles from downtown Newport. It has a commanding view of the Bay. Discussions in the *Newport Daily News* brought issues concerning the island to the citizens in the following reports in late 1880. The First report was December 6, 1880:

> THE TRAINING SCHOOL – Since the training school has been brought so prominently to our notice, it is well for us to know all the facts in regard to it in order to judge correctly of the present position of the matter and its bearing upon the action of our people.
>
> The general Government has for some years devoted three or four vessels to the use of boys who have been enlisted as "apprentices" in the navy. They are taught rudimentary English studies and practical seamanship and from time to time those most advanced are sent to sea on the regular ships of the navy where they are fully prepared to do services as petty officers. The system, which has long been in operation in England and France has proved to be an admirable and effective adjunct to the naval services and by reason of its successes in the crude manner in which it has been carried on so far, has led those most interested in the management of our navy to the conviction that the government should establish a permanent and central station and maintain an institution similar to those which have worked so well in Europe. In England there are over 3,000 boys and in France quite as many, who are under instruction in such an institution but on all our practice ships there are only 900 which number it is proposed to increase to 1,500. These boys are under constant discipline and at the recent naval review of Fortress Monroe, their appearance and general bearing most favorably impressed the President and other distinguished gentlemen present.
>
> *Where shall this school be established?* is a question that has arisen in Washington, and when it was first broached parties from Connecticut and New London, of course put forth New London because the Government already owns the land. About the same time, the General Assembly of Rhode Island unanimously passed a resolution appointing a committee of three with Gov. Van Zandt as chairman, to tender a site on Narragansett bay, such as should be selected by the Government.

Gov. Van Zandt went in person to Washington and communicated the action of the Assembly to the authorities but nothing was done for many months. Last spring, however, Congress appropriated $2,000 to be expended at New London and it was evident that one of two more appropriations like it spent there would make the plant large enough to prove an inducement to place the Training School there permanently, in spite of its natural disadvantages. Hence, in September last the Committee of the General Assembly addressed the Secretary of the Navy requesting him to send a board of officers to see if there was not a superior site for a training school in Narragansett Bay.

The board came here last week after examining the New London site and in company with our city officials visited Coasters Harbor Island, which everybody knows has unusual advantages for just such an institution. They also had under consideration two or three other sites on the Bay which could be passed upon by means of the charts with which they were provided. What their report will be no one can tell, as it must first be rendered to the Department, but in case it should recommend Coasters Harbor Island the definite and final action would be taken by Congress in deciding upon the location.

Strong influences will be brought to bear upon members of Congress by those interested in having the school at New London, on the Hudson, in the Chesapeake bay and at other places and we regret to say that in similar cases the result has often been decided by influence rather than the best interests of the Government. It is very certain that if we want the training school here we shall have to take prompt action to secure it.

The advantages accruing to the city would undoubtedly be very great as it would add to our population in winter about 2,000 persons. Of these probably 50 would be naval officers whose families would reside in the city. The direct expenditure would be about $500,000 per annum and the incidental amount very much more than that.

It has been stated that inasmuch as the General Assembly appointed a committee to tender a site and negotiate for its purchase after the government had selected one, the state would purchase Coasters Harbor Island of the city if that site was recommended. On the other hand it is said that the General Assembly of 1878, made no appropriation for that purpose and that in consequence the whole matter would come before the present General Assembly as new questions involving a large expenditure to be decided upon its merits. Some people claim that the honor of the state is involved by the resolutions passed in 1878, and to keep it untarnished the General Assembly must make an appropriation; but again it is said in reply that it takes something stronger that a resolution to bind the state, whose treasury is in the hands of a different set of men every year. We shall not further extend the pros and cons but venture to express the opinions that if Newport gets the advantage of the training school she will have to offer the location. Our people might as well meet the question now and decide it. "Shall the City Council be authorized to tender Coasters Harbor Island to the Government as a site for the Naval Training School?" The Council meets tomorrow night and can give every taxpayer a chance to express his opinion by a vote at an early day. This is the simplest, quickest and best way of deciding this important question and if the result should be in the negative that would end the matter; if in the affirmative it would remain for Congress to accept or not as was the case when we tendered the island for the use of the naval academy, but did not have to give it as stronger influences than ours returned that institution to Annapolis. Put this question before the people and it will be so thoroughly discussed in the News by correspondents to whom its columns are open whatever their views may be, or in public meetings, that in ten days it can be definitely settled one way or another and we can then devote our attention for the rest of the winter to other great questions like sewerage, public health, water, etc. and discuss to our heart's content. Gentlemen of the Council, give us the question.

December 9th from *The Daily Newport* reported:

> . . . the commission which had visited Coasters Harbor Island had unanimously recommended it as a suitable site for the United States Training School. It now remains for the citizens of Newport to decide . . .

On December 13th, again from *The Daily News:*

> It is said that the petition circulated on Thames Street was freely presented to all classes and that but three persons were found who did not give their signature.

On December 15th, a notice was posted in *The Daily News* stating:

> Important meeting of the City Council to take action upon the proposition to cede Coasters Harbor Island to the government.
>
> . . . President Franklin to read the petition which contained the names of upwards of three hundred taxpayers.
>
> The resolution amended read as follows: "Shall the City Council be authorized to give Coasters Harbor Island to the state of Rhode Island without consideration, to be by the state ceded to the United States for the location of the United States Training Station?" The aldermen voted to correct the ward lists on Monday next and to order an election for Wednesday, December 22nd.

There was a lot of discussion in the papers with columns and columns on the subject. The paper even printed the full report of the Naval Board on December 16th with the conclusion stating:

> . . . we would state that the Coasters Harbor Island possesses every requisite as a site for a naval training station, and we therefore earnestly recommend its acceptance. Respectfully submitted, Ned Earl English, Commodore U.S. Navy and Chief of Bureau of Equipment and Recruiting; Samuel R. Franklin, Captain U.S.N.; S.B.Luce, Captain U.S.N.; Ralph Chandler, Captain U.S.N.; F.E. Chadwick, Lieut. Com. U.S.N.

On December 22, 1880, as reported in the Daily News:

> The meeting at the Opera House last evening – a stormy night, but a large attendance – addresses by ex-Governor Van Zandt and Lt. Gov. Fay – Arguments in favor of the cession of Coasters Harbor Island.
>
> At eight o'clock the meeting was called to order by L.D. Davis who had been selected as chairman. The meeting had been called in accord with the grand American idea of government by the people.
>
> In conclusion the speaker asked for a large vote and a large majority, so that the question should be decided beyond a doubt.

Needless to say, the referendum passed by a large majority. The next step was up to the Rhode Island State Government, as it was against the Rhode Island State Constitution to allow a City or Town to cede property to the Federal Government and the City had to first give the Island to the State. The State of Rhode Island in turn transferred ownership to the United States Government on March 2, 1881. Two years later, in June of 1883, the Training Station was formerly opened by ther Secretary of the Navy William E. Chandler. The first commander was Commodore Stephen B. Luce who later became a Rear Admiral.

At first, trainees were housed onboard the U.S.S. *New Hampshire* which was docked at South Point.

Seamen drilled on shore, but lived on board ship. The double frigate had been brought to Newport in 1881. The ship was heated with steam and lighted by electricity. It accommodated about 500 apprentices who slept in hammocks. The *Richmond* replaced the *New Hampshire* after an outbreak of typhoid fever. The *New Hampshire* was taken to New London to be "frozen out" to remove any traces of the disease. Because the *Richmond* was a smaller ship, only a portion of the apprentices could sleep on board, and it was then that some were barracked on Coastal Harbor Island. The first recruit buildings were built on South Point in September of 1883: a drill hall and a gymnasium. Up until this time, with the exception of the typhus outbreak, apprentices lived on board ship, up until 1887. In 1894, the ship U.S.S. Constellation was assigned to the Newport Training Station until the close of World War II.

> The U.S.S. *Constellation*, attached to the Training Station, is a 36-gun frigate authorized by Congress in 1794. This full-rigged vessel, 161 feet in length and carrying a 167-foot mainmast, saw initial service in the unofficial war with France, 1798-1800, and next took part in the Mediterranean operations, 1801-1805, against Barbary pirates. In 1840, the *Constellation* was sent out to the Far East to protect United States interest in China. After long service in many parts of the world, it became a receiving ship, being stationed at Norfolk and Philadelphia, and then at Annapolis as a practice ship for midshipmen in the U.S. Naval Academy. One hundred years after her keel was laid, the historic frigate arrived at Newport.
>
> ~Louis Capellis
> *Rhode Island: A guide to the Smallest State*

Author Note

A word about captions to the postcards: Pertinent information will be listed in the following order as available:

1. In italics, the printed information from the picture side of the card
2. The approximate date of the card (often a postmark date)
3. In italics or quotes, written messages by the sender or printed data
4. Author's commentary

Chapter One
A New Century: 1900-1915

The New York Tribune was quoted in the Monthly journal *Our Naval Apprentice* of May 1902:

Growth of the Navy – The American Navy has undergone a remarkable transformation in the last generation. The "new navy" had its beginning in 1882, under Secretary of the Navy Hunt. Since the war with Spain the navy has risen by leaps and bounds, and has actually been almost quadrupled in the last four years. No such progress has been achieved by any other navy; in fact, the advance made by other naval powers sinks into insignificance beside the stupendous development in the United States. In 1880 the appropriations amounted to about $12,000,000; in 1887 there were about $15,000,000, and it had not passed much beyond $20,000,000, when in 1898, came the war with Spain. The present Congress is asked for about $100,000,000 for the navy, but a larger proportion of the amount is needed for construction of new vessels and the continuance of work on those under construction."

The Apprentice goes on to report:

Another Increase to Our Navy – In the naval appropriation for the next fiscal year money has been set aside for new ships – two battle ships of about 16,000 tons displacement, to cost, exclusive of armor and armament, $4,212,000 each; two armored cruisers of about 14,500 tons displacement, to cost, exclusive of armor and armament, $4,659,000 each, and two gunboats, of 1,000 tons, each to cost $382,000.

Extracts from annual report of Commandant of the Naval Training Station, Newport, R.I., for the year ending June 30, 1902. The number of apprentices under instruction during the year was 2,316. Of this number were: At the Station, June 30, 1901 – 958. Received between June 30, 1901 and July 1st, 1902 – 1358. Transferred between June 30, 1901 and July 1, 1902 – 1176. Discharged between June 30, 1901 and July 1, 1902 – 499. Deserted – 28. Dropped from roll, Bureau's order – 1. At Station, June 30, 1902 – 612."

The publication titled *Our Naval Apprentice* provided the following information:

During the year, recruits were transferred to the following ships: 446 to the U.S.S. *Monongahela*, 351 to the U.S.T.S. *Essex*, 302 to the U.S.T.S. *Hartford*, 5 to the U.S.S. *Alabama*, 4 to the U.S.R.S. *Columbia*, 2 to the Navy Yard, Wash. D.C., 2 to the U.S.F.S. *Kearsarge*, 1 to the U.S.S. *Cincinnati*, 1 to the U.S.S. *San Francisco*, 1 to the U.S.R.S. *Wabash*, 1 to the U.S.S. *Brooklyn*, 1 to the U.S.R.S. *Franklin*, 1 to the U.S.S. *Topeka*, 1 to the U.S.S. *Prairie*, 1 to the U.S.S. *Rainbow*, 1 to the U.S.S. *Alliance*, 1 to the U.S.S. *Buffalo*, 1 to the U.S.S. *Marietta*, 1 to the

U.S.S. *Chesapeake*, 1 to the U.S.S. *Massachusetts* and 1 to the U.S.S. *Olympia*. 5 recruits died while at the training station in 1901. The Commandant at the Newport Training Station in August of 1902 was Captain John J. Hunker.

The Naval Apprentice was printed monthly at 34 Clarke Street in Newport. Subscription to *The Apprentice* was $1.50 per year and was printed monthly. An advertisement at the back of *The Naval Apprentice* from the Newport One Price Clothing Company on Thames Street offered Chief Petty Officers' Uniforms for $10 and $12. Navy trade was solicited by local merchants.

The postcards in this chapter depict the early 20th century at the Newport Naval Training Station – the years 1900 to 1915.

Admiral Chas. M. Thomas, Commandant U.S. Training Station. Newport, R.I. Circa 1900. Thomas commanded the *Constellation* and was Commander of the N.T.S. July 27, 1904-July 6, 1905.

Battle Ship Rhode Island at Fore River Iron Works, QUINCY, Mass. January 13, 1904.

Repelling a Landing Party. U.S. Naval Training Station, Newport R.I. "Aren't we having a glorious day for Newport? Aug. 17, 1905. Belle"

Newport, R.I. Training Station. Camp Grounds. Circa 1905.

United States Naval Training Station, Newport, R.I. July 31, 1906.

Full Battalion Marching at Newport in Front of War College. Circa 1906.

"Fighting Bob", Admiral Robley D. Evans, U.S. Navy. September 15, 1910. Robley Dunglison Evans (18 August 1846-3 January 1912) was born in Virginia and served in the U.S. Navy from the Civil War to the Spanish-American War. In 1907-1908, he commanded the Great White Fleet. A frank and plain-spoken man who had a reputation for profanity, he was called to task by a pastor from Litchfield Connecticut in a letter to the *New York Times*.

U.S.S. *Virginia taken while at Anchor in Newport Harbor.* September 29, 1906.

Sham Battle, U.S. Naval Training Station, Newport, R.I. August 21, 1907.

Newport, RI Training Station Drill Room, Inspection Barracks B. August 21, 1909. "This is where the boys drill inside. Ralph"

Newport, R.I., Instruction on Jack Stay, Naval Training Station. May 21, 1907. "Why don't you write? Navy will go out of commission if you don't. I'll be home June 5 in the afternoon for an hour or so. Jack Far."

U.S. Training Ship "Constellation" Newport, R.I. Instruction on the different ropes. August 15, 1910. "Dear M – Package and letters received with pleasure – there is a big time here this week – so many ships coming in – four thousand sailor boys are to be ashore a day. It is pretty warm here. Write soon – love to you all. L."

Newport, R.I., Semaphore Drill Naval Training Station. September 23, 1907. (Handwriting on the front side) "Left Chicago two weeks ago, was down in Norfolk, Va. Stayed in New York two days. Have been here a week and is going to stay here three months / will then go aboard a ship / I was too tired of Minneapolis. Do you see me in this bunch? Address Thomas Houghton Naval Training Station / Barracks A, Room 5 Newport, R.I." (On the back side) "Newport, R.I. Miss Oby. I am now here in Newport Rhode Island in the Naval Training Station, have joined the U.S. Navy are going over on the Pacific Coast in December. Hope to get a letter soon, Yours forever. Thomas / Don't wait too long before you write"

U.S. Sailors Life "A Boxing Match" January 3, 1907. "We are on our way to Newport, R.I. for Torpedoes – we will not be there long, Harry"

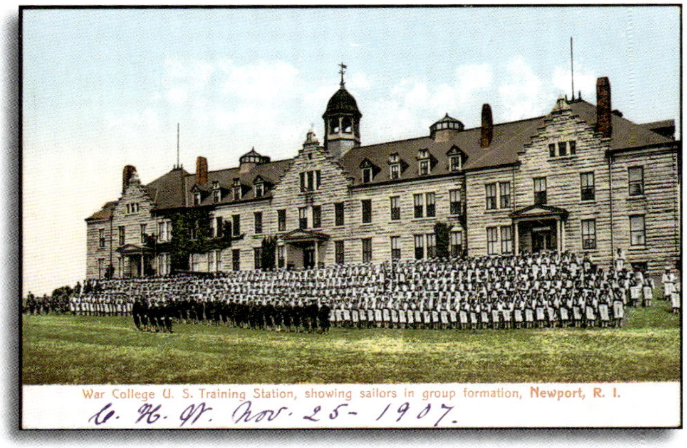

War College U.S. Training Station, showing sailors in group formation, Newport, R.I. Circa 1907.

New Government Landing Station, Newport, R.I. February 5, 1907. The row of buildings on the horizon is Long Wharf. The location of the Landing Station is approximately where the Newport Harbor Hotel and Marina is today.

Newport, R.I., Qualified Squad ready to leave General Service, Naval Training Station. Circa 1907.

Newport, R.I., Scrubbing Clothes, Naval Training Station. Circa 1907.

Newport, R.I., Boat Drill, Naval Training Station. Circa 1908. This postcard was addressed to: Mr. Archie F. Arnold, Portsmouth Navy Yard, Portsmouth, N.H. U.S.S. *Marietta*.

U.S. Receiving Ship "Reina Mercedes." Captured Spanish Vessel. May 27, 1908. The *Reina Mercedes* was involved with the Spanish American War and was purposefully sunk by Spain in Santiago Harbor in an attempt to block the harbor. The ship failed to block the harbor and the U.S. raised the *Mercedes* starting January 2, 1899. The ship was towed to Norfolk and Kittery for refitting. It was a non-self-propelled receiving ship and was towed to Newport in 1905, where she remained until 1912. The ship was decommissioned in November 6, 1957 and sold for scrap.

Newport, R.I. Training Station, showing Barracks B & Constellation. Circa 1908.

Training ships, "Reina Mercedes," "Cumberland" and "Constellation," Newport, R.I. Circa 1908.

Private photo postcard. December 8, 1908. "I just now received you letter. . . where I am standing in the picture is the end of the bridge. The bridge is about 30 or 40 feet long. Those are the lifeboats below me. Those instruments on the bridge are used for signaling at night. That is a telescope spy glass under my arm. Walter"

U.S. Battleship "Rhode Island" at anchor, Newport, R.I. Circa 1908. Launched, May 17, 1904. Part of the Great White Fleet. Decommissioned 1923 and sold for scrap.

Instruction in Signaling, Naval Training Station, Newport, R.I. August 9, 1909. "Dear Lydia: Arrived in Newport at 2:15 Monday morning and feeling fine. Address my mail U.S. Training Station, Newport, Rhode Island-Barracks A, Room 16. Otis"

Battery B Barracks, Newport RI. Circa 1909.

Newport, R.I. Submarines, Shark & Porpoise at Newport Training Station. Circa 1909.

Studio postcard taken at Juan Casablanca – No.10 Colonial Street, Newport, Rhode Island. Before WWI.

U.S. Training Ship "Constellation" in Newport, R.I. Instruction in pulling the Yards. July 5, 1909. "Dear Lester. I got here all right and like it fine. But you have to be up too early. We have to drill 5 days a week. We play baseball on Sunday. Sully."

U.S. Training Ship "Constellation" showing men scattered aloft, Training Station. Newport, R.I. Circa 1910. "Dear Brother – the reason I am sending this in a envelope is because I am sending a letter to ma and also to pa. This is a fine training ship. Your brother Ray."

Naval Apprentices Parade, Newport, R.I. Circa 1910. Up until the Training Station closed, Navy personnel marched in all Newport parades.

Boat Drill. U. S. Naval Training Station. Newport, RI. Series II. Circa 1910.

Toner Road. Camp Sadler. U. S. Naval Training Station. Newport, RI. Series II. Circa 1910.

U.S.S. Old Constellation and U.S.S. Boxer. U. S. Naval Training Station, Newport, RI. Series IV. Circa 1912.

Singing in Mass. U. S. Naval Training Station. Newport, RI. Series II. Circa 1910.

The 8th Regiment Camp. U. S. Naval Training Station, Newport, RI, Series IV. Circa 1912.

The 6th Regiment. U. S. Naval Training Station, Newport, RI. Series II. Circa 1910.

Salem Coaling at Bradford, Coaling Station. Newport, R.I. September 1, 1910.

Newport, R.I. Naval Training Station. The Training Ships. Postmarked: "U.S.S. Constellation April 19, 1911."

Ready for the Clam Bake at Newport, R.I. on Fleet Day. October 13, 1910. "Dear Doctor: I have inlisted in the U.S. Navy for a career. I would like to hear from you. My adddress is 1st Brigade, 3rd Division, U.S. Naval Training Station, Newport. I am in Newport tonight and thinking of you, thought I would send you this card. Walter Browne"

Naval Apprentices on Parade at Newport, R.I. October 19, 1911. These sailors march on the first block of Broadway. The buildings in the background are still standing today. The red home in the center is the Wanton Lyman Hazard home and is Newport's oldest standing residence.

Electric Studio Newport, R.I. Studio portrait postcard Circa 1910. This is a very happy young sailor. His name is Andy.

Constellation Guard on Board U.S. Training Ship "Constellation," U.S. Naval Training Station, Newport, R.I. November 12, 1912. "I am having a fine time but in the Navy now. I am well and hope you are well to and family to. From F.L."

April 4, 1913 Thames Street, Newport, R.I. The Electric Studio 162 Thames Street, Newport, R.I. and other cities. Four Navy buddies have their portrait taken at a downtown photo studio.

Naval Boys on Parade and Thames St. Newport, R.I. Circa 1913. Thames Street was Newport's main business street.

Naval Training Station, Drill Grounds, Newport, R.I. August 4, 1914.

Training Station, Boys on Parade Broadway, Newport, R.I. August 29, 1911. Marching across from Newport City Hall. Notice the trolly tracks in the middle of the street.

Photo by W.H. Mowrey 97 Thames Corner of Marlborough, Newport, R.I. Circa 1912. Signed "Fred, Bill, Phil." Studio Photographers had background images available specifically for Navy portraits, such as the one behind these sailors.

Juan Casablanca No. 10 Colonial Street, Newport, Rhode Island. Circa 1914.

Training Station Newport RI – Guns captured from the Spanish at Manilla. Circa 1915. Commodore George Dewey engaged and destroyed the Spanish Pacific Squadron in the Battle of Manila Bay. These guns were from one of the Spanish ships.

Brenton Reef Light Ship, Off Narragansett Bay, R.I. July 6, 1908.

Circa 1915. "Decoration Day. Made while marching down Broadway in NewPort." The trees in the upper center of the images show the location of Liberty Park. Some of these building are still standing today.

Naval Training Station, Newport, R.I. Circa 1915.

Apprentice Seamen at drill, U.S. Naval Training Station, Newport, R.I. September 28, 1912.

Physical Drill. U.S. Naval Training Station, Newport, R.I. Circa 1912.

The Yeoman School, U.S. Naval Training Station, Newport, R.I. Circa 1912.

Signal Drill Letter "U," U.S. Naval Training Station, Newport, R.I. July 15, 1912.

Newport, R.I. Training Station. Lessons in Alphabetical Drilling by Arms. (Handwriting on the front side) "This is part of B.C. One building is not shown but it is just like these two. They are all just alike inside but are for different divisions. The position these fellows are making is W. I will send Pete the semaphore and you and him can learn to signal together." October 17, 1910. (On the back side) "Hello Ethel; I got your postal yesterday. I looked quite natural so much that it made my homesick. I do this drill every day, it is a signal drill each movement means a letter and I can do it now just as fast as I can talk. My address is now Signal Division-Barracks C. Newport. W.P."

Navy Boys on Parade, Newport, R.I. June 25, 1913.

Newport, R.I. Barracks B,. Training Station. Circa 1912. Barracks B, built in 1904, was acquired by The War College in 1946 and renamed Sims Hall. It was used for the Navy Warfare Development Command.

Newport, R.I. Training Station—Group of Middies. Circa 1912. (On the back side) *Training Station—Group of Middies. The principal training station for enlisted men and for apprentice is at Newport. 'Middies' is an abbreviation of midshipmen, a title in the U.S. Navy substituted for naval cadet, and ranking next below ensign. The term is derived from the fact that in the British navy the cadets were given quarters amidships.*

Instruction in 3 inch Field Guns, Naval Training Station, Newport, R.I. November 7, 1912.

Newport, R.I., Company of Infantry under fire, Naval Training Station. Circa 1912.

Newport, R.I. Physical drill under arms, Naval Training Station. Circa 1912.

Newport, R.I., Instruction on Jack Stay, Naval Training Station. Circa 1907.

Newport, R.I., Training Station. Training with 3 inch field Guns getting in position. August 3, 1909.

Guard Mount Barracks "B" Naval Training Station Newport, R.I. October 13, 1913. "Dear Ray. How be you. Am feeling great just got in time about half past ten. Am feeling kind of hungry. I ain't going up to the barracks until to-night. Well be good from Elmer."

Newport, R.I. Torpedo Station. August 5, 1913. (On the back side) *The Torpedo Station is on Goat Island, and includes a school, maintained by the Government for experimentation with and manufacture of various explosives. A great number of torpedoes and electrical instruments are stored, and it is here that the naval officers go for practice during the summer months.*

Newport, R.I. Naval Training Station Barracks No. 1 and 2 Newport, R.I. September 10, 1915.

View of Harbor from Battery Park, showing Torpedo Station, Newport, R.I. July 21, 1913. "We are sitting in this park. I can see 12 big ships very near. The war ships have been here since Wed. We are going to take the Ocean Drive this P.M. Went to the beach Sat. Fannie"

Newport, R.I., Knotting and Splicing, Naval Training Station. Circa 1914.

Government Landing, Newport, R.I. Circa 1914. On the right side you can see one of the Fall River Steam Ship liners docked at Long Wharf.

Government Landing-Newport Waterfront. Circa 1915. The Navy had acquired Government Landing by 1913. Today, the site is occupied by Bowens and Bannisters Wharf, both filled with restaurants, shops, and pedestrian walking areas.

Souvenir Album – Naval Training Station Newport, R.I. War College, Newport, R.I. Addressed to: *Miss G. Tulin, c/o T. Doubing, Jackson Ave., Mineola. L.I. N. Y.* Postcard albums of the Training Station were popular offering up to a dozen postcards in booklet form. This album allowed you to either mail the entire booklet (for 3 cents) or break away individual postcards.

Naval Training Station, Newport, R.I. At the Saluting Battery, U.S. Naval Training Station. Another booklet with fold-out views of the Training Station.

The following represent a sample of battleships from The Great White Fleet in the early part of the 20th century.

U.S.S. Alabama. Launched May 18, 1898. Sponsored by Miss Mary Morgan. Displacement: 11,565 tons. Length: 374'10". Speed 16 knots. Compliment: 536. Decommissioned March 1, 1923.

U.S. Battleship Connecticut. August 19, 1910. Launched September 29, 1904. Sponsored by Miss Alice B. Welles. Sold for scrap November 1, 1923.

Battleship Kansas, 18.1 knots. Displacement 16,000 tons. Length 456′ 4″. Speed 16 knots. Compliment 880.

U.S.S. Michigan. First American Dreadnought Class, 1909.

U.S.Navy, Crusier "Topeka". (On the back side) *Crusier Topeka. Built at Kiel, Germany. 250 feet in length and 35 feet in breadth, and displacement about 2,400 tons. Speed, about 16 knots per hour. Horse-power, 2000. Cost, about $170,327. Armament made up of rapid-firing guns from 4-inch size down. Compliment, 21 officers and 131 men.*

U.S.S. Florida. Launched May 12, 1910. Sponsored by Mrs. Percival (Elizabeth Legore Fleming) Hamilton. Decommissioned in 1931 to comply with the terms of the London Naval Treaty and scrapped the following year.

U.S.S. South Carolina. August 2, 1917. (On the back side) *South Carolina Speed 18.86 knots. Displacement 16,000 tons. Length 450 feet. Horsepower 18,000. Armament: 8-12", under 4"-26, 2-21" torpedo tubes. Complement 49 officers, 756 men. Cost: $3,540,000.*

U.S.S. Wyoming. Lead ship of a class of two 26,000 ton battleships. Built in Philadelphia, PA and commissioned in September of 1912.

U.S.S. Indiana. (Handwriting on front side) *Dick we arrived at Norfolk this morning at 6 o'clock. Glad we are on land Mamma.* April 23, 1906. Authorized in 1890 and commissioned 5 years later. Small in size but built with heavy armor and ordinance. Designed for coastal defense. Partly responsible for the destruction of the Spanish destroyers *Pluton* and *Furor*. Sold for scrap in 1924.

Crusier Maryland in Dry Dock Charlestown, Mass. (Handwriting on front side) *I am glad you and grandma had such a pleasant visit. I thought of you that very rainy night.* March 1, 1908. Launched September 12, 1903. Displacement: 13, 680 tons. Length: 504'. Propulsion 2x vertical, inverted, triple expansion engines. 23,000 iph. 2 x screws. Sold for scrap Feb. 11, 1930.

U.S. Battleship Ohio. The third *Ohio*, laid down April 22, 1899 by Union Iron Works, San Francisco, CA. Launched May 18, 1901. Sponsored by Miss Helen Deschler. Captain Leavitt C. Logan was in Command. Designated flag ship of the Asiatic Fleet. Embarked for Manila in 1905 with Secretary of War William Howard Taft, including Miss Alice Roosevelt.

U.S. Armored Crusier Colorado. (Handwriting on front side bottom) *We arrived home Monday afternoon. Remember me to all I know. Write me your address when you go to Bethel and I will send the music. Gladys.* (Handwriting on front side upside down) *We all had a lovely time up there. Have you loaded any hay lately? Didn't we have sport that afternoon?* The second *Colorado* was launched April 25, 1903 by William Cramp and Sons Ship and Engine Building Co. Sponsored by Miss. C.M. Peabody. Captian D. Kennedy in command. Joined the Atlantic Fleet October 11, 1905. Scrapped October 2, 1930.

U.S. Monitor Purtian. Circa 1910.

U.S. Battleship Iowa. April 16, 1909. Launched March 28, 1896. Builder: William Cramp & Sons. Cost: $3,000,000.00. Sunk as a bombing target March 23, 1923.

U.S.S. New York. March 9, 1907. Laid down September 19, 1890. Launched December 2, 1891. Decommissioned April 29, 1933. Scuttled December 1941.

Chapter Two
The Great War and Beyond: 1914-1925

The Training Station at Newport saw a steady increase in personnel from the time of the Spanish-American War through World War One. Thousands of men arrived as civilians and left ready to serve the Navy and their country. Newport became the most important East Coast training facility, and by 1917, the Station could accommodate 2,000 sailors. Temporary barracks were built to accommodate the influx up to the War. Quite a few Navy personnel rented civilian quarters in the Point and Hill sections of Newport. With the increase in Navy recruits, Newport established programs including social, religious, and cultural events. An Army – Navy Y.M.C.A. was built in Washington Square in downtown Newport at a cost of $315,000 donated by Mrs. Thomas Emery of Cincinnati. Very popular with sailors, 5,735 crossed the threshold in a single day just before WWI. The Navy exercised wartime powers in March of 1918 to prohibit the sale or possession of alcohol within 5 miles of the Station. In April of 1919, several hundred Marines tried to "rescue" a fellow Marine who had been arrested by local authorities and the local press described the event as a "small Bolshevist riot."

An article printed in *Scientific American,* May 12, 1917, titled "The 12½ Minute Cafeteria" discussed how recruits at the Newport Naval Training Station were fed. Six messes were served to recruits starting with breakfast at 6:15 AM and 8:15 AM; dinner (our lunch) at 11:15 AM and 1:15 PM; supper at 4:15 PM and 6:15 PM.

> The two-hour interval between messes is necessary for the galley and servers to prepare the second mess of the same meal. In preparing the menu given some 2,100 loin pork chops were cut, 1,050 loaves of bread baked and five huge tubs of spuds.

In 12½ minutes, with the assistance of 21 servers, all 1,800 recruits were served. Two pork chops, two potatoes, tapioca pudding, and bread would be a common supper-time meal. The article included four photos showing kitchen staff and recruits chowing down.

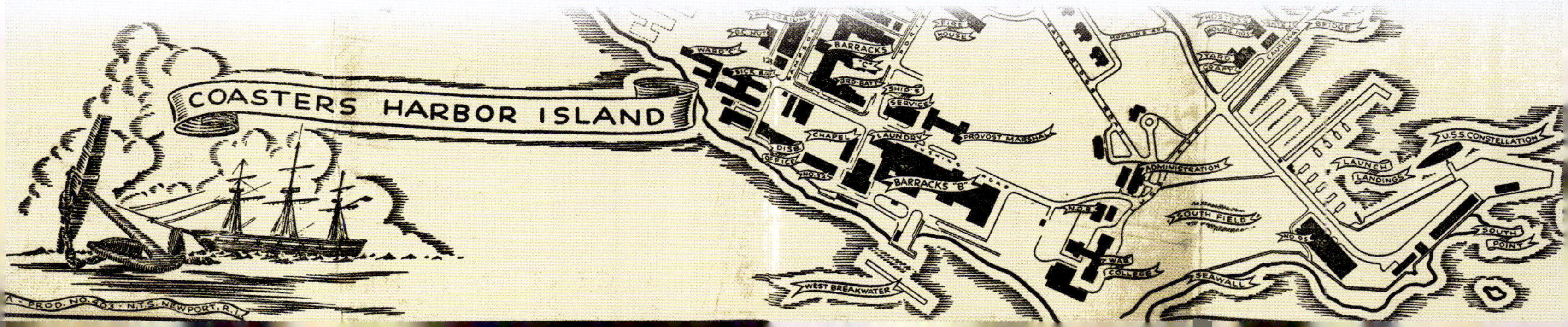

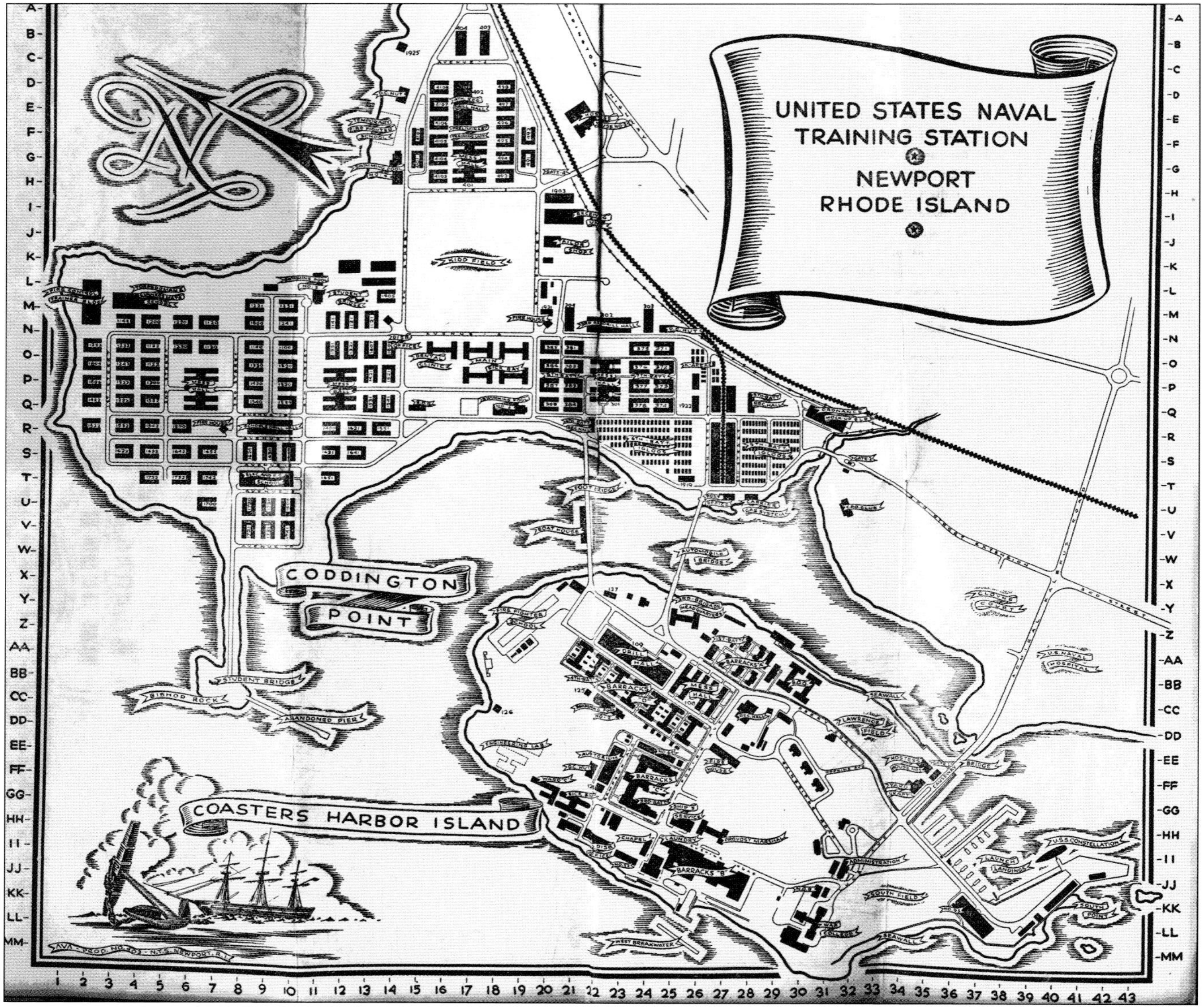

This document was given to each recruit with one side showing the map of the Training Station and the reverse showing a map of downtown Newport featuring churches, the Y.M.C.A., U.S.O. Club, beaches, and attractions.

Training Station showing Yeoman School, Newport, R.I. Circa 1914.

Signal Instruction, Training Station, Newport, R.I. Circa 1916.

Newport R.I. Visual Signaling. Circa 1915.

"Bag Inspection" Barracks "C" Naval Training Station, Newport, R.I. Circa 1916.

After a few months at the Training Station, Newport. Afternoon Parade. Circa 1916. This is a postcard sent to the family of a newly inducted sailor. The next postcard shows the back side.

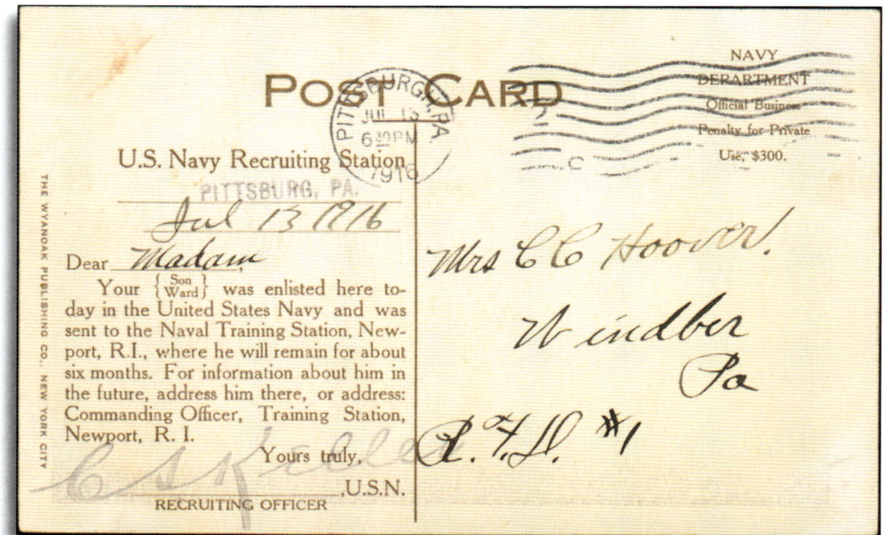

POST CARD

NAVY DEPARTMENT
Official Business
Penalty for Private
Use, $300.

PITTSBURGH, PA.
JUL 13
6:30 PM
1916

U.S. Navy Recruiting Station
PITTSBURG, PA.
Jul 13 1916

Dear Madam

Your {Son / Ward} was enlisted here to-day in the United States Navy and was sent to the Naval Training Station, Newport, R.I., where he will remain for about six months. For information about him in the future, address him there, or address: Commanding Officer, Training Station, Newport, R. I.

Yours truly,

C S Kelle, U.S.N.
RECRUITING OFFICER

Mrs C C Hoover.
Windber
Pa
R.F.D. #1

THE WYANOAK PUBLISHING CO. NEW YORK CITY

Back side of previous postcard sent from the Naval Recruiting Office to the family of the newly enlisted sailor. Circa 1916.

Bunker Hill Recruits just receiving Arms, U.S. Naval Training Station, Newport, R.I. October 5, 1916.

Circa 1916. Rugen Studio, Portraits, Photo Supplies, Electric Novelties, Artists Materials. 167 Thames Street, Newport, R.I. There were over a dozen photographic studios in Newport at this time and naval portraits were a significant trade to those businesses.

"Newport, R.I. Training Station.
R. Ferguson 1917"

Circa 1917. The Electric Studio 162 Thames Street, Newport, RI. Photographs of Quality Developing and Printing for Amateurs.

Captain's Inspection Naval Training Station Newport, R.I. Circa 1916. The Training ship Constellation is docked in the bay.

(Censored) Naval Training Station, Newport, R.I. Circa 1917. (On the back side) *Naval Recruits Trained at Newport Naval Training Station. Here is part of their course they need to train for. Each man claims life time of experience. Photo shows a new batch of recruits at their first mess, and they are hungry. The training course at the Newport Training Station fits the men for their place on Uncle Sam's ships.*

Interior K of C Hut. Naval Training Station, Newport, R.I. (Censored) Circa 1917. The notation "Censored" was used in war time to make sure that no critical logistic information was inadvertently revealed.

Company Barracks "B" U.S. Training Station, Newport, R.I. April 25, 1917. Sent from and written on the U.S.S. *Boxer*: "Well Lefty, Am having a fare time here. I haven't seen you here as yet. You want to come. Our company is going on a ship this week. Remember me to all the boys. Bert – My address is 3rd Division, 1st Company, Barracks 13, Newport Training Station"

German Submarine U. 53 at Newport, R.I. Britannia beherrscht die Wellen !! An deutscher Kraft wird dieser Rul zerschellen! January, 1917. U. 53 visited Newport October 7, 1916. It was a Type Mittel U U-boat of the Kaiserliche Marine, one of 6 forming the sub type 51 used during WWI. The commander was Hans Rose. As strange as it may seem, after visiting the Newport Port, the U. 53 proceeded to attack and subsequently sink Allied ships all while surrounded by American destroyers! Though many ships were sunk, no lives were lost with the American Navy rescuing all attacked passengers and crew. Outside of national boundaries, American ships had no right to interfere with the attacks.

Circa 1917. Group shot of sailors in front of Barracks No. 19, Newport Naval Training Station, Newport, RI.

Newport, R.I. Training Brig "Boxer" under sail, Naval Training Station. Circa 1916. The *Boxer* was the fourth of that name launched October 11, 1904 by the Portsmouth Navy Yard in Portsmouth, New Hampshire. Served as a training vessel in Newport until October 20, 1912, sent to the U.S. Naval Academy until June 25, 1914 and returned to Newport until May 14, 1920, when she was transferred to the Department of the Interior.

Circa 1917. Portrait of three sailors at the Newport Naval Training Station, Newport, RI.

Constellation Guard. U.S. Naval Training Station. Newport, R.I. Circa 1917.

"L. Hall Co. L. Bks 7. 2nd Naval Dist. Newport, R.I." Circa 1917.

Military Mass, Barracks B. Father Gleeson, Officiating. Circa 1917.

Circa 1918. Photo postcard, Newport Naval Training Station showing the Y.M.C.A. in the background.

U.S. Naval Training Station, Newport, R.I. In front of Commanders Headquarters. Circa 1919.

Barracks "C" Training Station, Newport, R.I. Circa 1919.

U.S. Naval Recruits in Training like their "chow". July 16, 1918. "7-16-18. This will give you an idea of the way we are fed. Only the place we go to is a big building-get your stuff as you go along-then pass on to a table and stand up and eat. Great life. Walter"

Telephone Exchange and Officers' Quarters, Newport, R.I. June 5, 1918.

Naval Training Station, Barracks C., Newport, R.I. August 23, 1918. "Very busy and tired, will write later. Had about 2 hrs. sleep last night, so I am going to make up for it now. Walt"

A Draft for Sea Duty, U.S. Naval Training Station, Newport, R.I. November 13, 1918.

U.S. Naval Training Station, Newport, R.I. Naval Recruits lined up for inspection. June 4, 1918.

U.S. Naval Training Station, Newport, R.I. Company on Parade Grounds. Circa 1919.

In front of Commanders Headquarters, U.S. Naval Training Station, Newport, R.I. August 4, 1918.

Receiving Barracks, U.S. Naval Training Station, Newport, R.I. Circa 1918.

Naval Training Station, Newport, R.I. September 13, 1919. The Drexel Hut, Y.M.C.A. *'The Drexel Hut' Y.M.C.A. is unquestionably the most popular spot at the station, with great writing facilities, a bank, library, athletic floor, and auditorium where every man can find just the proper thing to occupy his leisure time.*

U.S. Naval Hospital, Newport, R.I. May 15, 1918. "Dear Mother: received your letter and will answer soon. I have just passed my examinations and am being transferred to the base hospital at Newport. N.H. Wiley HAIC-U.S. Naval Hospital, Newport"

Auxiliary Yacht "Aloha". Circa 1918. Built as a private yacht for Commodore Arthur Curtis James (1866-1941). James was an industrialist and yachtsman who summered in Newport. The U.S. Navy acquired the *Aloha* from James April 22, 1917 for use as a patrol vessel during the war. The ship was commissioned June 5, 1917 as USS *Aloha*; returned to James January 29, 1919. James toured around the world on the *Aloha* in 1921 and 1922. She was the only bark-rigged yacht of her size afloat at that time. The *Aloha* was scrapped in 1938.

Drexel Hut, Y.M.C.A., U.S. Naval Training Station, Newport, R.I. Circa 1919

The Hospital Corps Training School, U.S. Naval Training Station, Newport, R.I. Circa 1919.

The Naval Hospital, U.S. Naval Training Station, Newport, R.I. Circa 1919.

The Yeoman School, U.S. Naval Training Station, Newport, R.I. Circa 1919

Officers' Quarters, U.S. Naval Training Station, Newport, R.I. Circa 1919.

The Commandant's Quarters, 2nd Naval District. U.S. Naval Training Station. Newport, R.I. Circa 1919.

Torpedo Station, Newport, R.I. Circa August 17, 1918. "August 26, Moving to our new quarters across the bridge. Will write later. Wait until I give you my new address before you write. L. My new address is 2nd Naval District. Barracks # 9, Company C. Newport, R.I."

U.S. Naval Training Station, Newport, R.I. Hospital School. Circa 1919.

K. of C. Hut. Naval Training Station, Newport, R.I. Circa 1919.

For you and Uncle Sam. Circa April 22, 1918. "Dear Friend, I will give you my address 2nd Co. U.S. Naval Training Station, Newport, Rhode Island. Write to me, I will write a letter next time from your friend Otto Easterly. Answer soon"

Scrubbing Clothing – Training Station Newport, R.I. August 13, 1919.

The Receiving Barracks. U.S. Naval Training Station, Newport, R.I. Circa 1919.

Visitors' Day at Drill Office. U.S. Naval Training Station, Newport, R.I. Circa 1919.

Recruits leaving Receiving Station. Newport, R.I. April 8, 1922. "I guess you will be surprised to hear from me in this part of the country, but I surely have been having the time of my life. I like the Navy life, am going to take up Radio work as soon as I receive a second class seamanship which will be about 2 months. Well if you can't be good, be careful. Your W. Sharp"

U.S. Naval Training Station, Newport, R.I. Barracks C. September 11, 1923.

Where the Tars are made. May 31, 1922.

Naval Training Station, Newport, R.I. Circa 1922. *The War College, as seen from the administration building, one of the many fine buildings at the station.* The War College was established in Newport October 6, 1884.

Chapter Three
Between the Wars: 1926-1940

During this era in Newport, the Training Station proved to play a critical part of the local economy. At least one third of the new homes built in Newport were for employees of various Navy installations, including the Training Station. Locals, whom at one time worked seasonally for the Bellevue Avenue elite and found themselves laid off for much of the year, found full-time employment with the Navy. In 1936, the Navy's budget on Aquidneck Island was $7,500,000 and employed over 3,500 civilians. Three quarters of those lived in Newport and of that fifty percent were homeowners.

Instruction at the Training Station had advanced to the level that, in 1931, the Station produced a *Course of Recruit Instruction* booklet. The 86-page booklet detailed day by day the activities that would turn a civilian into a "Tar." Recruit training took twelve weeks and included instruction in these (among other) subjects: scrubbing clothes, cleaning stations, bag inspection, boats, compass, knots and splices, anchors and ground tackle, rigging, sails and rigging, deck gear, power boats, painting, splicing, whipping, and cutting. Each of the many categories were detailed with Knots and Splices given as an example:

> Overhand; square; marline hitch; clove hitch; rolling hitch; how made; uses of; material. Review small stuff, spun yarn, ratline stuff, seizing stuff. Short, eye and long splices.

Recreation and Liberty included leave to visit Newport, which included the downtown Y.M.C.A., the Seaman's Church Institute, and movies. Writing home, club rooms, games, reading, study, library, and athletic exercises were also encouraged. Bag Inspection reviewed the recruits' abilities in "laying out, folding, marking, stowing, hammock and bedding." Having another person's clothing in possession was considered a serious offense. The Commanding Officer, as stated in this booklet dated 2 November 1931, was G. J. Rowcliff, Captain.

In another publication titled *Newport Recruit,* dated December 12, 1936, an article titled "Year's Last Dance" showed the social relationship between Newport and The Training Station.

> The last Station Dance of 1936 will be held in the Auditorium on Saturday, 19 December, starting at 2000. The orchestra has prepared several special numbers and effects in keeping with the holiday season. Invitations to several hundred young ladies in Newport and vicinity for this dance already have been addressed and will be placed in the mail on Monday 14 December.

Another article of note titled "Bravo! 6th Company" informs the reader:

> The sum of forty-four dollars, ninety-five cents ($44.95) was turned in by Company Six, which graduated from the U.S. Naval Training Station on November 14, 1936. This money pays for one of the newly installed stained glass windows in the Station Chapel, and the presentation inscription will be inserted upon arrival from the factory in Providence. Such inscriptions are baked upon the glass and will be permanent records of worth while g fts toward the Training Station Chapel, doubly appreciated when one considers the small pay during training period.

Motion pictures shown during this week (December 12, 1936) were a 10-reel film, *Page Miss Glory,* staring Marion Davies and Dick Powell, and a 7-reel film, *Hot Tip,* staring James Gleason and Zazu Pitts.

From 1933 to 1935, the Station did not accept recruits.

United States Naval Training Station Newport, R.I. Circa 1926. Series III.

Approach to Administration Building. U.S. Naval Training Station, Newport, R.I. Circa 1926. "Dear Mother: have a look at the place where I am staying, of course it is not summertime here but it's pretty warm once in a while. I wish the shower baths were as warm as the picture looks. See the little car in the picture, that's the one that brought me here. (Ladis E. Zack).

Boat Drill, U.S. Naval Training Station, Newport, R.I. Circa 1926. "Do not send me any money yet as we are going to move soon, please write and tell me if you have received this card as I am very anxious. I am getting a thirteen day leave and it will be my last chance to come home for four years." (Ladis E. Zack)

Returning from Drill, U.S. Naval Training Station, Newport, R.I. Circa 1926. "They will not let a fellow go home if he hasn't got enough money for a round trip ticket, that's why I want you to send $56.00 but don't send it until I write again." (Ladis E. Zack)

Exhibition Drill, U.S. Naval Training Station, Newport, R.I. Circa 1926. "When I received them two letters from home I felt like a king. Tell Evelyn I enjoyed her letter very much. I hope George gets that Ford but I hope he don't get his head knocked off." (Ladis E. Zack)

Mass Singing, U.S. Naval Training Station, Newport, R.I. Circa 1926. "Dear Mother – I'm not going to take up engineering. I'm getting a transfer to the band where they will send me to a music school for one year. I will not get on a boat until I have completed the music course." (Ladis E. Zack)

Main Entrance to Station over Causeway, U.S. Naval Training Station, Newport, R.I. Circa 1926. "This music course will not cost me a cent and it will make me a professional cornet player – here's hoping it does – Har!Har!" (Ladis E. Zack)

Captain's Inspection, U.S. Naval Training Station, Newport, R.I. Circa 1926. "Just as long as I receive a letter from home I will not be homesick." (Ladis E. Zack)

Boat Drill, U.S. Naval Training Station, Newport, R.I. Circa 1926. "See the black boat in the picture, that is the Constellation — the oldest ship in the American Navy –150 years old. I worked on it already. I had my picture taken Sunday. I hope I will get them pretty soon. I am going to order a dozen which will cost me $4.00. I don't get paid for the first month here." (Ladis E. Zack)

Putting clothes on the line to dry, U.S. Naval Training Station, Newport, R.I. Circa 1926. "Tell Mrs. Norton I might like the Navy alright but I don't like the idea of washing my own clothes or dishes. From the whole room there are 107 fellows in this company." (Ladis E. Zack)

11th Regiment Camp, U.S. Naval Training Station, Newport, R.I. Circa 1926. "Well I can't think of anything more to say so I guess I will close. I have to start washing my under ware leggings and jumper now." (Ladis E. Zack)

Regimental swimming party, U.S. Naval Training Station, Newport, R.I. (Handwriting on front side) *Ladis E. Zak Joined the Navy Feb. 2nd, 1926.* Circa 1926. "Well good by Mother and take care of yourself as I want to see you when I get home. I don't gamble because we are not allowed to have dice or cards around and I wouldn't gamble any way. There are kids here 13 years old and it's awful." "Gave by your son Lad – Ladis E. Zak N.T.S. Company 12, Room 20, Newport, Rhode Island."

Harbor View Showing Fort Adams, Newport, R.I. June 23, 1924. *Harbor View showing Fort Adams which commands the entire Newport Harbor and is located at its entrance garrisoned by three companies of Coast Artillery.*

United States Naval Reserves Drilling at Newport Naval Training Station. Circa 1927.

The New Brigade Headquarters, Naval Training Station, Newport, R.I. Circa 1927. *Naval Training Station, Newport, R.I. The new Brigade Headquarters, an example of the Jackies skill, built by enlisted men.*

U.S. Naval Airship. Circa 1928.

Naval Training Station, Newport, R.I. Circa 1930. *Naval Training Station, Newport, R.I. The "Grinder," where many "happy" hours are spent. Showing the hospital Corps School, Gordon Hall and the Detention Barracks.*

Swedish Drill. U.S. Naval Training Station, Newport, R.I. Circa 1936. "Letter coming. Very busy and feeling fine. Never better in my life. Matt"

Bag Inspection, U.S. Naval Training Station, Newport, R.I. Circa 1936. "Les: This is No. 1. I am having a swell time. I don't know how the beer is but the drilling is about the same as what we had in the Hose Co. Vic"

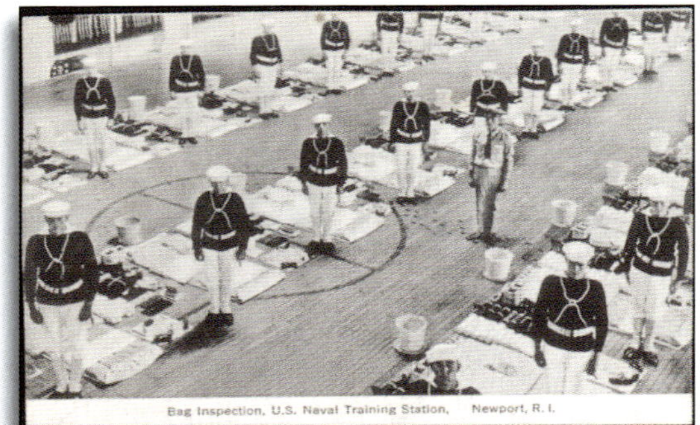

Yeoman School, U.S. Naval Training Station. Newport, R.I. Circa 1936.

Apprentice Seamen on Review. U.S. Naval Training Station. Newport, R.I. Circa 1936.

Toner Road, Camp Sadler, U.S. Naval Training Station. Newport, R.I. Circa 1936. "This is a line of shacks which are our homes. The one with the X is the rear end of the one I live in."

Barracks "B", Naval Training Station, Newport, R.I. Circa 1937. "Dear old Willie (Wilhelmina), I'm sorry to have to send you only a little card but you just wait a while and I will make it up to you on a letter. Well anyway I sure am having a grand time here and wish you were here too. The only thing I miss up here are the good old girls. As you see I must remain in detention for three weeks and then am I going to make up for lost time. Well old Pal of most Pals, I must be closing. By by Jessy"

Training Ship "Constellation," Newport, R.I. Circa 1937.

Dormitory Barracks "B", U.S. Naval Training Station, Newport, R.I. October 10, 1936. "This card will give you an idea of how we sleep and if we should walk in my sleep I'd break my neck. Ray"

Army and Navy Masonic Service Center. Masonic Temple. Newport, Rhode Island. Circa 1937. *Conducted by The Masonic Service Association of the United States.*

Bag inspection aboard ship. Naval Training Station, Newport, R.I. May 17, 1927. "Dear Mom: Came over here tonight to see "doin's" Meg Reg and two officers from Squantum. They flew down here yesterday and are going out at daybreak tomorrow to scout for "Black" fleet. Wallace"

Ships Company, U.S. Training Ship Constellation. Newport, R.I. Circa 1937.

Naval Training Station, Newport, R.I. April 19, 1927. *Washday at the Naval Training Station. Why are the uniforms of our sailors so clean and spiffy looking? The answer is found in this picture where we see a group of jackies at work in the laundry of the Naval Training Station. Taking the black spots out of a certain country also lies in the "cleaning up" task of these boys.*

General View Mess Hall and Barracks. Naval Training Station, Newport, R.I. September 5, 1926.

Barracks "B" with view of harbor. U.S. Naval Training Station, Newport, R.I. Circa 1939.

Boat Drill. U.S. Naval Training Station, Newport, R.I. Circa 1939.

U.S. Naval Torpedo Station. Newport, Rhode Island. Circa 1939.

"Constellation," Newport, R.I. August 1, 1939.

U.S. Naval War College, Newport, Rhode Island. Circa 1939.

U.S. Naval Training Station, Newport, R.I. Circa 1939.

Chapter Four
World War II to 1953

The eminent possibility of war called for resumed training and the Station expanded, accelerating its capacity to the level so that at the height of World War Two over 16,000 recruits were receiving training on base at the same time. The Commander of the Training Station during WWII (May 31, 1941-March 6, 1945) was Captain C. W. Magruder. The *Newport Recruit,* an 8-page newsletter (dated September 25, 1943) for the Station, detailed articles about the War, Liberty Log listings, theater listings, which included *Lady of Burlesque,* staring Barbara Stanwick and *Bombardier,* staring Randolph Scott. An interesting report titled "Refund of Taxes" stated:

> Service men with average incomes discovered last week, when they studied their estimated income and tax for 1943, that they would not have to pay any tax for this year. In most cases their 1942 taxes will be cancelled completely, as well. It now appears that the policy of the Internal Revenue Bureau will be to credit any overpayment, first, toward the 1943 tax and, second, toward the 1944 tax.
>
> ~Reported by D.H. Kelsey, CSp (T)

An official Navy training manual titled *Useful Information For Recruits* published in 1942-1943 started with the introductory statement:

> The duty of the Naval Training Station is to train recruits from civil life to become "Man-O-War's Men," and to inculcate in them discipline, spirit, and efficiency in the performance of Naval duties. The primary function is the preparation of men, as thoroughly and as quickly as possible, for duty in ships of the U.S. Fleet.

The duration of stay was stated:

> The period of recruit training is regulated by the Bureau of Naval Personnel, Navy Department, Washington, D.C., and may vary between three and twelve weeks, depending upon personnel requirements of the Fleet at the time.

On the subject of "Use of Intoxicating Liquors":

> It is a violation of a City of Newport, RI Ordinance to sell intoxicating liquors to recruits in the City of Newport.

Interesting, because "Blood Alley" in downtown Newport was a common bar-hopping destination for the Navy, and the Shore Patrol made regular runs in this area, returning drunk sailors back to the base. Concerning leave and liberty, the pamphlet stated:

The City of Newport, in addition to being the location of many important naval activities, is a city possessing several points of historical interest, and is also a famed summer resort with many fine beaches. Entertainment is provided by motion picture theaters, roller skating rinks, the Army and Navy YMCA, the USO, and other service clubs. There are several restaurants which made a specialty of serving lobster and other sea-food delicacies from waters off the New England Coast.

This pamphlet was the property of Recruit William Laing Troup, Jr. He was in Company 969, Barracks 498 and his Company Commander was C.B.M. Hawks. During the war, training was suspended in 1944 and was re-opened again in 1950. The base was officially closed as a training base for recruits January 12, 1952. Very few buildings have survived since the Training Station days and a recruit from WWII would be amazed at the changes. What did survive were the deeds and actions of the thousands of Navy personnel who passed through the Training Station and went on to serve the Navy and their country. The citizens of the U.S. and their allies will forever be in debt to you for your service.

Army and Navy Masonic Service Center—Masonic Temple, Newport, Rhode Island. Circa 1941.

Walk, Dexter Training Grounds. Circa 1941.

Library, U.S. Naval Training Station, Newport, R.I. Circa 1941. "I haven't been here. Won't for 3 weeks, it's between 'C' & 'B' Barracks."

Naval War College, Newport, R.I. Circa 1941.

U.S. Naval Training Station, Newport, Rhode Island. Circa 1941.

Barracks "B" with view of harbor. U.S. Naval Training Station, Newport, R.I. Circa 1941.

Aerial View U.S. Naval Training Station, Newport, R.I. Circa 1941.

Aerial View Fort Adams, Newport, R.I. Circa 1940.

U.S. Naval Training Station Barracks, Newport, R.I. December 11, 1942. "Dear Friend. Just a line to let you know I haven't forgot you. Well I'm in the Navy. Wish you a Merry Xmas and Happy New Year. A friend Eddie Lane-CO-356. Write Soon"

U.S. Naval War College, Newport, Rhode Island. Circa 1940.

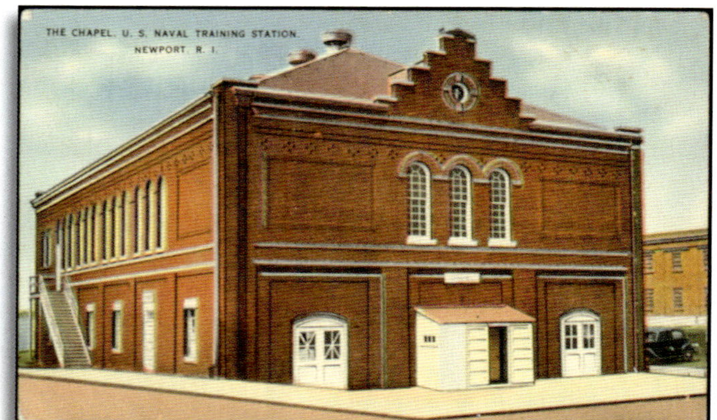

The Chapel. U.S. Naval Training Station, Newport, R.I. August 21, 1941. "Hi Frank. Well I turned out to be a nice guy. Now go to church every Sunday. Haven't left the barracks since I got here but will Saturday. I guess the town will soon have beginning sailors. Frank."

U.S. Naval Training Station. War College and Drill Field, Newport, Rhode Island. Circa 1941.

Anchors of "Constellation," Newport, R.I. January 20, 1943. "Hello Cecile: Sorry I could not see you before I left, am having a very nice time since meeting fellows from all parts of the U.S. Give my best regards to your folks, your boy-friend and Aunt & Grandmother. Eddie K."

U.S. Naval Training Station, Newport, Rhode Island. August 14, 1942.

Government Landing and Newport Harbor, Newport, R.I. February 18, 1944.

Telephone Center. Naval Operating Base. Newport, R.I. September 24, 1944. "This is where we make phone calls but the picture only shows about 1/3 of the room. There is 19 booths in the place so can usually get a call through pretty fast. Joseph Savarll, Co 1392 NTS Newport, R.I." (postage was marked free)

U.S.S. Brownson DD-868. Circa 1946. *A unit of the U.S. Atlantic Fleet home-ported at Newport, R.I. on Narragansett Bay.* Launched July 7, 1945 and commissioned November 17, 1945. Decommissioned September 30, 1976. Displacement 3,460 tons.

Destroyers Gearing, Zellars, Norris and Charles P. Cecil. Circa 1950. *A unit of the U.S. Atlantic Fleet home-ported at Newport, R.I. on Narragansett Bay.*

U.S.S. Fred T. Berry DD-858. Circa 1949. *A unit of the U.S. Atlantic Fleet home-ported at Newport, R.I. on Narragansett Bay.* Launched January 28, 1945, commissioned May 12, 1945. Home port Newport, arriving September 11, 1949. Stricken from the Naval Vessel Register September 15, 1970.

U.S.S. Myles C. Fox DD-829. Circa 1949. *A unit of the U.S. Atlantic Fleet home-ported at Newport, R.I. on Narragansett Bay.* Launched August 14, 1945, commissioned March 20, 1945. Arrived in Newport May 23, 1949 serving in the North Atlantic for a year. Decommissioned October 1, 1979.

U.S.S. Ingraham DD-694. Circa 1945. *A unit of the U.S. Atlantic Fleet home-ported at Newport, R.I. on Narragansett Bay.* Launched January 16, 1944. Commissioned March 10, 1944. Displacement 3,218 Tons. Decommissioned July 16, 1971.

U.S.S. Massey DD-778. Circa 1946. *A unit of the U.S. Atlantic Fleet home-ported at Newport, R.I. on Narragansett Bay.* Launched August 19, 1944. Commissioned November 24, 1944. Displacement 3,218 tons. Decommissioned 1969.

U.S.S. Vesole DD-878. Circa 1946. *A unit of the U.S. Atlantic Fleet home-ported at Newport, R.I. on Narragansett Bay.* Launched December 29, 1944. Commissioned April 23, 1945. Displacement 3460 tons. Decommissioned and stricken December 1, 1976.

U.S.S. Charles P. Cecil DD-835. Circa 1946. *A unit of the U.S. Atlantic Fleet home-ported at Newport, R.I. on Narragansett Bay.* Launched April 22, 1945. Commissioned June 29, 1945. Reclassified March 18, 1949 and sent to Newport to be part of the Atlantic Squadron. Displacement 3,460 tons. Decommissioned October 1, 1979.

U.S.S. Joseph P. Kennedy, Jr. DD-850. Circa 1946. *A unit of the U.S. Atlantic Fleet home-ported at Newport, R.I. on Narragansett Bay.* Launched July 26, 1945. Commissioned December 15, 1945. Displacement 2,425 tons. Home port Newport. Steamed into Washington, D.C. for the inauguration of John F. Kennedy, brother of "her" namesake. President Kennedy and Jacqueline watched the America's Cup race on board the *USS Kennedy, Jr.* September 1962. Decommissioned 1973 and used today as a part of the Battleship Cove Museum in Fall River, Massachusetts. Open to the public.

U.S.S. Glennon DD-840. Circa 1946. *A unit of the U.S. Atlantic Fleet home-ported at Newport, R.I. on Narragansett Bay.* Launched July 14, 1945. Commissioned October 4, 1945. Displacement 3,460 tons. Overhauled in Newport, 1946. Decommissioned October 1, 1976.

U.S.S. Lloyd Thomas DD-764. Circa 1950. *A unit of the U.S. Atlantic Fleet home-ported at Newport, R.I. on Narragansett Bay.* Launched October 5, 1945. Commissioned March 21, 1947. Converted in 1949 to a hunter-killer destroyer and joined the Atlantic Fleet at Newport. Re-designated *DDE-764* March 4, 1950. Decommissioned October 12, 1973.

Souvenir Folder of U.S. Naval Training Station, Newport, R.I. Circa 1945. (Handwriting on front side) Mrs. Mabel Seavy. 19 Bay Street. Box 148. Belfast, Maine. (On the back side) *The United States Naval Training Station is the birthplace of the naval training system of this country. It is located on Coasters Harbor Island and was established in 1868 at the avocation of Lieutenant Stephen B. Luce, later a Rear Admiral, for the American training of enlisted men. The station now has well over 100 large and impressive buildings.*

U.S.S. Norris, DD-859. Circa 1950. *A unit of the U.S. Atlantic Fleet home-ported at Newport, R.I. on Narragansett Bay.* Launched February 25, 1945. Commissioned June 9, 1945. Joined the Atlantic Fleet in Newport after extensive alterations to increase her anti-submarine capabilities. Reclassified as a Destroyer Escort on March 4, 1950. Decommissioned December 4, 1970.

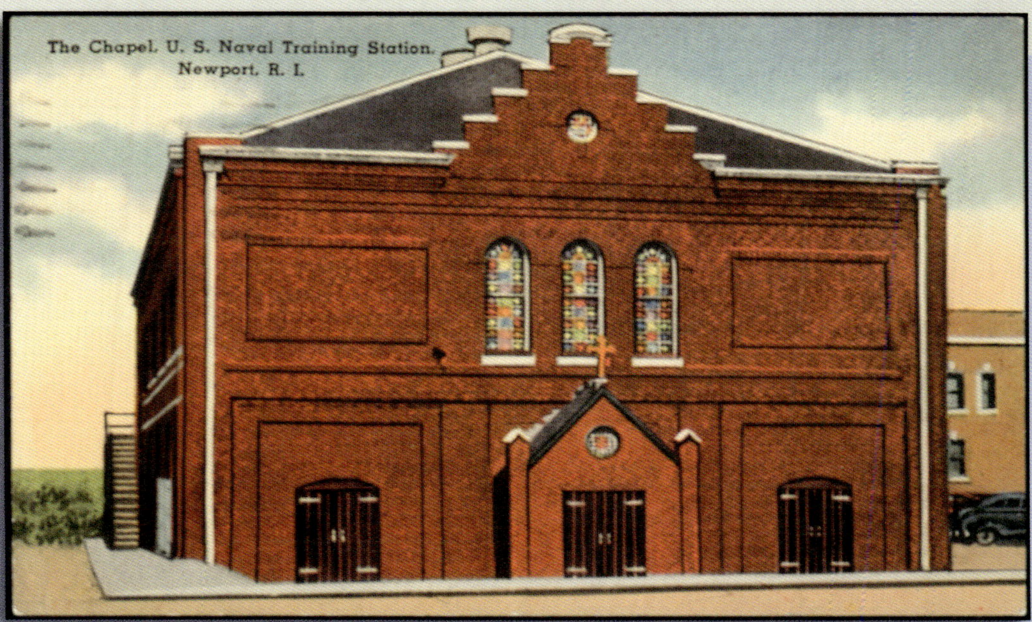

The Chapel. U.S. Naval Training Station, Newport, R.I. September 18, 1943. The area above the stamp box on the back side of the postcard has the written word "free."

Chapel, U.S. Naval Training Station, Newport, R.I. July 19, 1948.

Post 1950. Destroyer Piers – U.S. Naval Station, Newport, R.I.

Bibliography

Cappelli, Louis W. *Rhode Island a Guide to the Smallest State*. Cambridge, Massachusetts: The Riverside Press. 1937.

Gorski, Chaplin V.J., *Newport Recruit*. Newport, Rhode Island: NTS Newport, RI. I-12-12-36-1400.

Hattendorf, John B., B. Mitchell Simpson III, John R. Wadleigh. *Sailors and Scholars The Centennial History of the U.S. Naval War College*. Newport, Rhode Island: Naval War College Press. 1984.

Jackson, John E, Jondavid Duvall, Kimberly Rhoades. *Naval War College Illustrated History and Guide.* Printed in the United States of America by U.S. Government Printing Office. 2nd ed. , April 2010.

Mason, George C. *Newport Illustrated*. Newport Rhode Island: C.E. Hammett Jr. Publisher, 1854.

Munn, Charles Allen. *Scientific American*. New York, New York: Munn & Co., Inc. 1917.

Recruit Training Command. *The Helm*. New York, New York: Yearbooks Publishing Co. Inc.1951.

Rowcliff, G.J. Captain. *Course of Recruit Instruction.* Newport, Rhode Island: N.T.S. Newport – 11-30-1200. 1931.

Wyld, Lionel D. *The Naval War College: Images of America*. Charleston, SC: Arcadia Publishing. 1999.